EXTREME CAREERS

DEMOLITION EXPERTS

Life Blowing Things Up

Mark Beyer

the rosen publishing group's
rosen central

Published in 2001 by The Rosen Publishing Group, Inc.
29 East 21st Street, New York, NY 10010

Library of Congress Cataloging-in-Publication Data

Beyer, Mark.
Demolition experts: life blowing things up / by Mark Beyer.— 1st ed.
p. cm. — (Extreme careers)
Includes bibliographical references and index.
ISBN 0-8239-3365-2
1. Wrecking—Juvenile literature. 2. Wrecking—Vocational guidance—Juvenile literature. 3. Blasting—Juvenile literature. [1. Explosives industry—Vocational guidance. 2. Vocational guidance.] I. Title. II. Series.
TH447 .B48 2001
690'.26—dc21

00-012747

Manufactured in the United States of America

Contents

Fire!

Surrounded by an excited crowd of people, you stand a few blocks from the old hotel, waiting. The crumbling, ten-story building hasn't been used for years, and it's scheduled to be torn down today. Over the past few weeks, you've noticed crews of men and women working at the site to prepare the building for demolition. You can't believe that today is the day. Suddenly, a horn sounds. Somebody in the crowd tells you that it's six minutes away from blast time.

To keep the crowds far away from the building, police officers and firefighters stand guard on the street. The crowd is buzzing, and you can hear shouting near the hotel. The demolition

4

experts are ready to hit the button that will start the explosions. Waiting for the final horn, you feel like your heart will pound out of your chest.

There it is! The horn blows one last time, long and loud. Only one minute to go. A silence falls over the crowd. That minute seems to take an hour. You wonder who's more excited about the explosion: the demolition experts or you.

Finally, a man's voice comes through a loud-speaker. "Ten . . . nine . . . eight . . . seven . . . six . . . " The crowd counts along with him. After he gets to "one" he yells, "Fire!" Someone has pushed the button. From this distance, the exploding dynamite sounds like loud pops. Smoke flies from the hotel. You feel the rumbling blasts shake the ground beneath your feet.

All at once the far side of the building begins to fall. The near side of the building follows it. A huge cloud of dust rushes upward from the ground. The top of the building disappears into the dust cloud. Suddenly, the echoes from the explosions and the falling chunks of concrete stop. There is silence. The crowd erupts in a loud cheer.

A crowd of curious onlookers watches the demolition of the Travelers building in Boston, Massachusetts.

Learning to blow up buildings, bridges, ships, and other structures is much more complicated than one might think. Demolition crews don't just run through buildings tossing dynamite around. Using explosives to demolish things takes years of training and experience. A demolition expert lives a life filled with hard work and danger. The fun is the excitement of watching things blow up and fall down.

Why Blow Things Up to Knock Them Down?

Buildings and other structures are destroyed for many reasons. Some are unsafe, perhaps due to age, or because they weren't built well originally. Others are hazards to the environment. Still others are simply taking up space that could be used for something better. When it is decided that a structure needs to come down, it's time to call in the demolition experts.

When people think of demolition jobs, they may think of bulldozers and wrecking balls. The wrecking ball swings against a brick building and knocks huge chunks of it to the ground. With a mighty push, the bulldozer cleans up the rubble, which is then hauled away by trucks.

Many demolition jobs are still done using the wrecking ball and bulldozer. Today, however, some larger buildings, bridges, and smokestacks are too big to be knocked down with a wrecking ball. For those jobs, experts use explosives like dynamite. Some explosive-demolition jobs knock down twenty-story buildings or 200-foot-tall smokestacks. Using a wrecking ball on these structures would take too long and be overly dangerous.

There are also other kinds of demolitions that you may not even know about. For instance, what happens when a massive ship is stranded and sinking in the ocean? Or huge rocks and trees are blocking a highway? What about when criminals or terrorists use explosives to threaten the lives of others? It's likely that demolition experts would be called in to solve each of these problems. If you think that demolition experts are only trained to blast buildings, you will be surprised to learn what else they can do!

Explosive Demolitions

Modern explosive demolitions began with the U.S. Army, Navy, and Marines during World War II (1939–1945). These branches of the military used

Demolition Experts: Life Blowing Things Up

explosives such as dynamite and nitroglycerin to destroy enemy bridges, ships, weapon factories, and railroads. They also used explosives to blast through rock and soil to build canals, bridges, and airport runways. It was important for them to learn how much dynamite was needed to knock out posts that held up a structure and where to place dynamite to destroy an enemy target quickly. Knowing how to destroy any type of structure helped the Allies win the war.

After the war, the United States experienced a building boom. Returning soldiers and sailors with demolition experience possessed useful skills and easily found work with construction companies. They knew that if a building could be blasted with dynamite to knock it down, a lot of time and money

A U. S. Army demolition crew watches a dynamite explosion during World War II.

would be saved. Eventually, these demolition experts started their own companies. They traveled to different cities across the United States, knocking down old buildings to make way for new structures.

A Family Business

When their businesses grew rapidly, explosive-demolition experts didn't know where to turn for extra help. They couldn't recruit students from universities, since at the time no schools taught students how to use explosives to blast structures. The men who started their own demolitions businesses had to hire apprentices and train them to work with explosives. To protect themselves and other people during a demolition, also called a shoot, the apprentices were taught safety rules. They also studied architecture and engineering, because they needed to know how structures were built in order to figure out how to bring them down.

Only a few dozen people around the country knew how to use dynamite and other explosives to knock down structures. The demolitions field was so exciting and exotic that jobs were filmed and

A Forest Service Emergency

It's no secret that the U.S. Forest Service employs demolition experts to destroy diseased stumps and rocks that need to be moved for road construction. But did you know that these demolition experts are also called upon to destroy dead animals? The rotting carcasses can poison the water supply. They also attract coyotes and bears near trails or populated areas, endangering humans. Many of these animals have become contaminated or are too large to be buried. That's when the Forest Service uses explosives to blast the dead animals into pieces.

televised to the American public.

By the 1960s, because of great demand, demolition companies were working all over the world. The wives, sons, and daughters of demolition experts helped run the businesses. This is how demolition businesses are still run today. Many war veterans have turned their businesses over to sons and daughters, who have learned the art of explosive demolitions themselves. They carry on what their parents have taught them, and keep up on the latest technologies and safety practices. Today's demolition jobs are larger, trickier, and more dangerous than ever.

Knocking Down a Building

If you ask explosive-demolition experts how they go about blowing up buildings, they will all tell you the same thing. Blasting a structure depends on a combination of science, experience, and gut feeling. The science part of explosive demolitions comes from a combination of physics and engineering: knowing how to use explosives and understanding what makes a building stand up.

Experience comes from doing the job over and over. Gut feeling is something that every demolition expert counts on when it comes to knowing how to blast a building. Before any dynamite is set, demolition experts can visualize how a structure will fall. If all goes well, a controlled blast will knock the building down exactly the way the demolition expert predicted.

Demolition experts discuss the best approach to blowing up a building they have just examined.

Experts call a demolition job a "shoot." When they "blow the knees out" from a building or structure, they shoot it. All demolition experts shoot structures the same way. It's called the implosion method. They want to blast just enough of a structure so that it becomes unstable and gravity's forces pull it down in the exact place it stood. This method sounds easy enough. The work needed to get to the point of blasting, however, takes a lot of time and planning.

Visiting the Demolition Site

Demolition experts have many things to think about as they begin each job. When hired for a project, they first examine the building to be demolished. They need to know how the building is constructed in order to determine how best to collapse it.

Buildings are made in different ways. Some are constructed from steel beams and brick. Others are built from reinforced concrete (heavy iron bars set into the concrete make it stronger). Still other buildings are a combination of steel beams, reinforced concrete, and brick. Demolition experts have a plan to knock down each type of building.

The demolition expert must also consider other factors. The height of a building affects how it will fall. The area (the ground space that the building stands on) is important, too. The placement of the building is also a factor in knowing how to implode it. A structure with nothing around it is much easier to shoot than one surrounded by other buildings.

Sometimes weakening a building means cutting it in half. Dykon, Inc. decided that cutting a concrete building in Oklahoma would make it fall in on itself more easily. Dykon's experts broke apart floors with jackhammers and cut steel bars using welding torches. Parts of the outside walls were broken down with sledgehammers. When they were finished, one could look through the building from top to bottom. When the dynamite was exploded, the building folded inward and dropped to the ground.

Step One: Making the Building Weak

Every building that needs to be demolished may not be a crumbling mess. A building can be considered unsafe simply because it's old or run down. Even if it is ready to be demolished, it may still be strong enough to stand on its own. After all, buildings are designed to be strong and sturdy. Experts say that it is easier to build one than to tear it down.

Some buildings are even demolished when

they're still in good condition. In these cases, they are demolished not because they pose a safety risk, but because someone has decided to use the land for some other purpose—a different kind of building, or maybe a parking lot.

If the building being demolished is strong enough to resist blasting, the demolition experts face a threat to their safety. For example, if only half of the building falls, they must then tear down the remaining half, which can be very dangerous. The ideal situation is

A high-rise beam welder uses a welding torch to weaken the structure of a building that is being prepared for demolition.

Demolition workers often use jackhammers to undermine the structure of a building before blowing it up.

for the building to fall completely in on itself. The best way to ensure that this happens is to begin the job by weakening the structure.

Before placing explosives, demolition experts use drills, jackhammers, and welding torches to weaken buildings. They knock down walls, cut through steel beams, and drill huge holes in concrete posts. Each of these jobs makes the building a little bit more unstable than it was. Once unstable, the building only needs a little help from dynamite to knock it down. As it falls, the weakened building easily breaks apart.

Weakening a building takes weeks. It is hard and dirty work. It is also one of the most important steps in demolishing a building.

Step Two: Packing the Explosives

Once the building has been weakened, it's time to place explosives in key positions. Demolition experts use their knowledge of engineering when placing explosives. They understand which parts of a building support the structure and which parts, when blasted, will cause it to fall.

Each floor of a building is usually supported by concrete posts or steel beams. Demolition experts pack explosives into every post and beam on the lower two or three floors. When the dynamite explodes, the floors and ceilings fall down. As these lower floors collapse, gravity brings down the rest of the building. The building crumbles into large pieces of rubble as it hits the ground.

Demolition experts test a beam of a building's lower
floor prior to packing it with explosives.

A building may also be supported by H-beams—beams made of steel and formed into the shape of the letter *H*. This shape makes the H-beam so strong that demolition experts need a special type of explosive to destroy it.

Using dynamite to blast H-beams would not be safe or effective. The best way to knock down an H-beam is to cut it in half. The question is, How do you cut a steel beam? Welders cannot burn cuts into beams inside a building because they would be crushed when the beams fell. The answer is to direct a blast in such a way that the beam is sliced in half.

Demolition experts cut square holes at the same spot on each H-beam. Linear shaped charges (explosive devices that use a specific shape to create massive force in a concentrated area) are placed through these holes and against the beam so they will explode outward on a slant. Like dynamite, shaped charges are detonated (set off) in order. A row of beams will explode at the same time. As the series of explosions cuts the beams, and gravity takes over, the building begins to crumble on top of itself.

Dynamite

Although demolition experts use more than twenty different types of explosives to do their jobs, dynamite is most commonly used. A dry chemical powder that releases energy when it reacts to fire, a stick of dynamite is more than 100 times as powerful as a firecracker. Demolition experts like to use dynamite in their work because they know how much power each stick has. This knowledge gives them a better idea of how much to use. Dynamite is also easy to work with: It can be broken in half and pounded into place without exploding!

Step Three: Timing is Everything

It takes a lot of work to knock down a building, but how do crews perform the job with such exact skill? Demolition experts are able to control the timing of explosive blasts. Some go off right away, while others are delayed. Without the proper series of delayed blasts, an explosive demolition can send a building toppling over to crush another building, a street, or power and telephone wires.

Using delays assures that the building will drop straight down.

Built into the blasting caps that ignite the explosives, delays help demolition experts choose which parts of a building will begin to fall first. Making certain parts of a building collapse first allows gravity to do most of the work.

In order to make a building implode, or fall in on itself, its center must begin to fall first. This means the posts and beams in the middle of the building must be blasted first. Then the series of blasts moves outward toward the sides of the building. By the time the last blasts destroy the outer beams or posts, the middle of the building is falling to the ground. Now gravity can take over and pull the outside of the building toward the center. As this happens, the building settles in on itself and leaves almost no rubble outside its own footprint (the space that the building takes up on the ground).

After the Show

If the shoot goes well, there is only the footprint left to clean up. Explosive demolition experts do not clean up the mess they leave. Their job is complete when the building is safely brought down. Demolition experts check that everything about the shoot has succeeded. Then it's on to the next demolition.

A different type of demolition company cleans up the rubble. Now the bulldozers, cranes, and dump trucks arrive. They are paid to take away the rubble and ready the site for the new building that will rise on the same spot.

Careers in Demolition

Demolition companies are responsible for blowing up bridges, ships, launch towers, even soccer stadiums. There are all kinds of things that need to be demolished, and many types of organizations that need experts to do the work. If you're interested in a career in demolition, there are many ways to put your expertise to work.

Blowing Up Bridges

Just as buildings are made from different materials, so too are bridges. Most bridges are made of concrete or steel or a combination of both. Demolition experts

Demolition crews weaken the old Sailboat Bridge over Grand Lake in Oklahoma on February 8, 2000 before setting the explosives.

say that most bridges have one thing in common: They are built the same basic way.

Bridges often have huge pedestals (posts) to hold up the bridge floor as it spans a river or canyon. Short bridges have only two pedestals, one for each end of the bridge. Longer bridges can have three, four, or more.

Taking apart a bridge is more dangerous than building it. The bridge becomes weak as each piece is taken away. This is why demolition experts are hired to knock the bridge down using explosives. In one mighty blast, any bridge can be knocked down in just a few seconds.

Concrete Bridges

Concrete bridges are blasted almost the same way as concrete buildings. The bridge is first weakened to prepare it for blasting. Holes are drilled into the concrete, including where the bridge floor meets the pedestal. Blasting caps are set into the dynamite and packed into the holes. Copper wires lead to the blasting box hundreds of feet away from the bridge.

Demolition experts want to make sure that the spot where the floor connects to each pedestal is destroyed.

When the button is pushed, the dynamite detonates in a series of explosions which move from one side of the bridge to the other. One after another the floor sections break from the pedestals and drop into the water. Only the pedestals remain standing. They are too big and thick to be blasted safely using dynamite. Such a blast would send chunks of concrete all over. The pedestals will be destroyed using wrecking balls and giant jackhammers. Then the cleanup crews lift the concrete out of the water to be taken to the dump.

Steel Bridges

Steel bridges are shot using linear-shaped charges. Demolition experts place shaped charges in just the right places so as few explosives as possible are used. Experts only want to use what is needed, and no more.

The pedestals of some steel bridges are made from concrete, while others are steel H-beams. Bridges with concrete pedestals are blasted where the steel floor meets the pedestal. Bridges supported by H-beams are cut with shaped charges. When the blasting box sends its electric charge to the explosives, they cut the

H-beams at the connection. The steel floor drops into the water like a book drops to the floor.

However, bridges using H-beams aren't necessarily easier to destroy. Their pedestals, unlike those of concrete bridges, need to be knocked down right along with their floor. Demolition experts attach shaped charges to the connections between the bridge floor and the pedestals, and also near the bottom of each pedestal. The charges are blasted in a delayed series. First the bridge floor is cut. A half-second later the pedestals are cut. The sections of the

Close Call

In 1961 Jim Redyke of Dykon, Inc., was hired to demolish a smokestack in South Africa. Redyke planned to shoot the smokestack so that it would fall away from other buildings. This method is typical for all demolition projects.

When the dynamite exploded, the smokestack crumbled down on itself instead of falling over like a tree, as Redyke had planned. The concrete used to build the smokestack had been so poor in quality that the blast cracked it apart all the way up to the top! Luckily, no one was hurt.

Spectators watch the demolition of the ASARCO smokestack in Tacoma, Washington in January 1993.

bridge floor drop first. Then the pedestals topple over onto the floor sections. Everything sits in the river for the cleanup crews.

Silos and Smokestacks

Farm silos are built to last. They need to be strong enough to hold a lot of grain and withstand high winds and powerful storms. Built from concrete with steel bars (rebar) running the length of the structure, silos are difficult to knock down. Demolition experts need to use the right amount of dynamite so that a silo will fall exactly where planned.

As with all demolition projects, silos are first weakened in order to make the blasting easier. Jackhammers take down parts of the bottom silo wall. They expose the rebar used in the silo construction. Then holes are drilled to hold the dynamite.

Knocking down a silo is much like chopping down a tree. The holes are drilled in a series that wraps halfway around the silo. The idea is to blast out a huge wedge near the ground. After the blast, gravity takes over. The weight of the silo causes it to fall over where the wedge was blasted. The silo crushes itself into large chunks when it hits the ground. This makes it easier to clean up after the blast.

Made of metal, concrete, or brick, smokestacks are blasted like farm silos. The difference is that smokestacks are much taller than silos, which makes shooting them especially tricky. Demolition experts need to plan exactly where the smokestack will fall. If they make a mistake, the falling smoke-stack can damage power cables, telephone lines, or another building.

Bomb Investigations

Police departments use demolition experts to investigate bomb blasts. When an explosion occurs, the police need to understand what was used to make the bomb and how the bomb did its damage. Demolition experts can sift through the rubble and identify pieces that tell them many things about the blast.

Many of these demolition experts learn their skills while serving in the military. The U.S. armed forces

A police bomb-disposal specialist shows a
fellow officer how to defuse a bomb.

A Miner's Best Friend

Mining is well known to be dangerous business. Demolition experts who work with mining companies are often at the mercy of falling objects and toxic fumes.

Experts estimate that in the next few years, giant robotic devices will be laying explosives, going underground after blasting, and performing other duties that prove dangerous to humans working in the mining industry.

Safety isn't the only benefit. One robot has the potential to be 4 percent more efficient than human miners. That efficiency gain could save the coal mining industry $280 million a year!

use weapons and explosives more than any other organization in the world. Military demolition teams learn to use the most advanced explosives and are taught by the most knowledgeable experts.

Demolition experts who remain in the military are often sent to war-torn nations to investigate or defuse bombs. The Federal Bureau of Investigation (FBI) and Bureau of Alcohol, Tobacco and Firearms (ATF) also have trained explosives experts on their staffs. What you may see as a crumbled building is evidence to them. You see, bomb blasts explode a certain way depending on what kind of explosive was used. The switches and wires used to detonate the explosive can tell an investigator if an expert or an amateur made the bomb. Investigators are even able to tell if the bomb was made in a foreign country and brought into the United States. How do they do this?

Bomb investigators study criminal groups who use bombs to terrorize people. They know what kinds of bombs are used most often. Since governments control the sale of ingredients to make bombs, criminals have limited legal access to explosives. Investigators can track U.S. sales of explosives and detonator products. They also work with foreign police agencies to learn about new explosive products in other countries. Before an investigator comes to a bomb investigation site, he or she already knows what clues to watch for when figuring out who committed the crime.

Demolitions at Sea

Demolition experts have teamed with environmentalists to tackle one of the seacoasts' worst problems. Because of shipwrecks, oil leaks, and pollution, a

U.S. Navy demolition experts curbed an environmental disaster in 1999 when they set fire to the *New Carissa*, burning off more than 200,000 gallons of oil.

great deal of the sea life along America's coasts has been destroyed. To bring back the fish, plants, and other creatures that have died off, demolition experts have been hired to blast and sink old ships off the coasts of Florida, California, New England, and the Carolinas. Algae collects on the sunken ships and attracts fish. Thanks to demolition experts, these rusty old ships have become sunken treasure for millions of fish, helping to bring back sea life to coastal areas.

In February 1999, the oil tanker *New Carissa* ran aground near Coos Bay, Oregon, and spilled 140,000 gallons of oil onto the nearby coast. The beaches and wildlife areas were covered in thick, gooey oil. An approaching storm threatened to tear the ship apart and spill an additional 260,000 gallons of oil in the water. Tugboats could not pull the tanker away from shore. The only plan that could save the water and beaches from further damage was to set fire to the ship and burn off the oil. The U.S. Coast Guard called in U.S. Navy demolition experts.

The navy demolition experts rigged the *New Carissa* with 400 pounds of explosives and 600 gallons of gasoline. As the crew worked, the storm raged against the *New Carissa.* The navy experts finished the job and left the ship, then set off the explosives by remote control. The ship exploded into a huge orange ball of flame. More than 200,000 gallons of oil burned before the storm split the ship in half. Coast Guard Captain Mike Hall said that the demolition experts did the best thing possible for the environment. "Every gallon that is burned in the fire," said Hall, "is a gallon that is not on the shore of the coast of Oregon."

Demolition Safety

Every aspect of a demolition project is controlled. That is, the demolition expert works to make sure each step is carefully planned. Planning each step means making everything safe. The demolition expert makes sure the building is a safe place for the crew to work. He or she checks that the plan to knock down the building is safe for the area and also makes sure that the explosives are used properly for a safe shoot. When each step has been completed, it's time to blast.

The Shoot Area

Demolition experts look carefully at the building and area where they are going to work. The building is the project, but the area around it is just as important. Demolition experts consider many factors. Is the target building close to other buildings? If so, they have to make sure those buildings are not damaged during the blast. How many people will come out to see the blast? There are always going to be people watching on blast day. Demolitions companies work with local fire and police departments to see that the audience will be a safe distance from the blast. When is it finally safe to blast? A system of warning sirens and checkpoints alerts the experts that it's blast time.

Damage Control

Demolition experts are hired to destroy one or more buildings, called targets. That means they have to be careful about other buildings close to their targets. Often, a target building will be standing only twenty feet away from another building. This is when the demolition experts' skills are tested most.

As noted earlier, demolition experts can knock down a building on top of its footprint. Most implosion blasts leave only a few bricks on the sidewalk next to the demolished building! This is how every controlled blast should work.

When detonated, dynamite shoots concrete and other debris every which way. Sometimes these flying missiles will travel 200 feet from the blast. Demolition companies must take all measures necessary to protect houses and other buildings from these missiles.

Neighborhood Homes

It's not always large buildings that need to be protected on a demolition site. Sometimes the surrounding buildings are people's homes. Homes have windows, wooden doors, and aluminum siding that can damage easily. There are several ways to protect homes from flying debris

Demolition companies use a special geotextile fabric to protect homes and buildings. This fabric is made of thick, strong, woven material so that flying debris cannot punch holes through it. Demolition experts drape geotextile fabric over the sides of homes and buildings that face the blast site.

Dust from the Seattle Kingdome implosion settles on the neighborhoods surrounding the demolition site.

Unprotected windows can be broken during blasts. But flying debris is not the only part of a blast that can break windows. The loud sounds that explosives make send out blast waves. These airwaves vibrate windows until they crack. Geotextile fabric muffles the blast waves to protect windows from vibrating and cracking.

Sometimes entire walls are built to give added protection to homes and buildings. Cranes stack large steel containers to form a barrier against the blast. These walls can be twenty or thirty feet high. They help protect homes against sound waves and flying rocks.

Crowd Control

Having an audience watch the building or structure come tumbling down is part of the excitement for demolition experts. They enjoy hearing a crowd cheer after a successful shoot. However, crowds must be controlled. People often don't understand that getting too close to a blast can be harmful to them as well as their homes.

Demolition companies work with local police and fire departments to control crowds. Streets all around the blast site are blocked off from traffic. Wooden

Demolition crews usually create an off-limits zone around the demolition site to prevent injuries to onlookers.

barricades are set up to keep people far away. Crowds are kept at least 500 feet from the blast site. Even at this distance, crowds have a great view of the shoot.

Controlling Explosives

Demolition experts take many steps to control the debris produced when dynamite explodes. Besides protecting people and buildings, they also need to protect the relay switches that lead to all the blasting caps. Without protection, the switches can be damaged. If they fail to work, the building might not fall in the right spot. Once all the explosives are set into concrete or placed against H-beams, they are wrapped. Proper wrapping ensures that very little debris can escape.

Wrapping Dynamite

Many steps are taken to wrap each piece of explosive. The process is different for concrete posts than for steel H-beams.

Dangerous Business

Does dismantling buildings that house nuclear reactors sound safe? Demolitions company Controlled Demolition Group managed to make the job in Berkeley, England, trouble free. For safety reasons, Controlled Demolition Group was hired to reduce the height of the buildings by 19 meters. Working on the buildings was dangerous business because the demolition posed the risk of damaging the reactors themselves. The crew had to be extra careful, and extra prepared, to ensure that all safety and security standards were maintained.

Copper wires leading to blasting caps are usually bundled and tied together to protect them from flying rocks.

After each concrete post is packed and wired with dynamite, a large piece of clay is stuck into each drill hole. The clay plugs the hole and helps control the blast inward. Next, demolition experts wrap each post with chain-link fencing to control large chunks of debris. Finally, each post is wrapped in geotextile fabric to restrain small rocks from flying outside the building.

Because linear-shaped charges do not create much flying debris, H-beams have a simple plywood box strapped around them. Demolition experts wrap geotextile fabric around the plywood box as an added protection.

Wires and Relay Switches

Not every copper wire leading to blasting caps can be protected. Those wires are bundled and tied together for safety. When faced with flying rocks, they are stronger as a thick bunch than as single wires. Each relay switch, however, needs to be protected from flying rocks. Demolition experts use a simple method to protect relay switches. They cover the switches with bricks or chunks of concrete. This

Test Blast

Using the right amount of dynamite to blow out a concrete post is important for a successful shoot. Demolition experts like to make a test blast to see how their explosives will work. Demolition experts wire a few posts in different parts of the building. They also wrap the posts for safety to check their damage control wraps. When everything is set, they shoot the posts. If the posts crumble as planned, they know that they've used the right amount of dynamite. Then they can pack and wire all their explosives.

provides enough protection against flying rocks as the timed explosions go off one after another.

Blast Day

Blast day is what every demolition expert lives for. It's show time! Everyone on the project is excited. On the morning of blast day, the demolition experts check every wire and switch. They have to make sure that every explosive is set to go off when the button is pushed. This process can take hours. They don't want anything to be out of place. If something goes wrong, the blast might not work right. Worse, someone could get hurt.

Firefighters and police officers take to the streets. People come out to watch the blast. Streets are blocked and businesses are closed. It's almost like a celebration!

Radio Teams

While the explosive connections are checked, the area is blocked off for safety. The demolition crew

The demolition of a famous site attracts a lot of press coverage.

A demolition expert displays the blasting box as he explains how a sports arena will be blown up.

places people at checkpoints around the building. They are in constant radio contact with the command post that will hit the blast button. The checkpoints talk with the police and firefighters. They make sure that everyone is far enough away from the blast and that the site is safe. As they make their checks, a series of sirens sound to warn people and the demolition crew that blast time is near.

Blast Signal Codes

Every shoot has a series of warnings. The first warning sounds two long blasts of a horn or siren. This signals that six minutes remain until blasting. The checkpoint crews talk with the crowd-control people. The command post attaches the wires to the blasting box. A safety switch covers the blast button on the blasting box.

After five minutes, the checkpoints call in "all clear" from their radios. One long blast from a horn or siren sounds across the area. This is the final warning. It's one minute to blast time! The crowd buzzes with anticipation. Everyone looks toward the wired building.

With the push of a single button, the Seattle Kingdome begins to implode.

The command post counts the final ten seconds out loud. Then "Fire!" is called out. The button is pushed and the series of explosions begins. One section at a time, the building begins to fall. In about ten seconds it's all over. All that remains of the building is a cloud of dust.

When the dust settles, two short blasts from a horn or siren call the "all clear" signal. This means that it is safe for the demolition crew to walk on the site and check their work.

Celebration Time

Demolition experts have a lot to celebrate after each shoot. They sometimes have worked a month or more to arrive at this point. The shoot was a success. The building fell like a house of cards. Debris sits in the building's footprint. No damage was done to any surrounding buildings. This is how every

The Seattle Kingdome crumbles into a cloud of dust in about ten seconds.

shoot should be. Now it's time to go on to the next project. What do demolition experts think of their job? To them, it's a blast!

Glossary

blasting caps   Small explosive charges stuck into dynamite sticks to make them explode.

charge   An explosive.

debris   Rubble or wreckage.

defuse   To prevent an explosive from exploding.

delay   Allows blasts to explode at different times.

demolition   Destroying buildings and other structures.

detonate   To set off an explosion.

dynamite   A dry chemical wrapped in paper to form sticks used in explosive demolitions.

explosives   Dry or liquid chemicals that release huge amounts of energy when detonated.

footprint   The area where the building once stood.

geotextile   Synthetic material that retains its structure against explosions.

Demolition Experts: Life Blowing Things Up

H-beam A steel support post formed into the letter *H* for added strength.

implosion To make a structure fall in on itself.

linear-shaped charge Explosive device that concentrates massive force in a specified area.

pedestals Pillars or posts supporting the floor of most bridges.

reinforced concrete Concrete that has metal bars inside it for added strength.

rebar Thick metal bars used in reinforced concrete.

shoot To demolish a structure using explosives.

For More Information

European Demolition Association
Euclideslaan 2
P.O. Box 8138
NL-3503 RC Utrecht
Netherlands
+31 30 689 8905
Web site: http://www.eda-demolition.com

Institute of Explosives Engineers
Centenary Business Centre
Hammond Close
Attleborough Fields
Nuneaton
Warwickshire CV11 6R4

Demolition Experts: Life Blowing Things Up

United Kingdom
+44 24 7635 0846
Web site: http://www.iexpe.org

International Society of Explosives Engineers
29100 Aurora Road
Cleveland, OH 44139-1800
(440) 349-4004
Web site: http://www.isee.org

National Association of Demolition Contractors
16 N. Franklin Street, Suite 203
Doylestown, PA 18901-3536
(800) 541-2412
(215) 348-4949
Web site: http://www.demolitionassociation.com

National Fire Prevention Association (NFPA)
1 Batterymarch Park
P.O. Box 9101
Quincy, MA 02269-9101
(617) 770-3000
Web site: http://www.nfpa.org

Videos

What a Blast! 4 tapes, Brent Blanchard Productions. Unapix Home Entertainment, 1999.

Web Sites

Due to the changing nature of Internet l nks, the Rosen Publishing Group, Inc., has developed an online list of Web sites related to the subject of this book. This site is updated regularly. Please use this link to access the list:

http://www.rosenlinks.com/ec/deex

For Further Reading

Chudley, Roy. *Construction Technology: Tunnelling, Demolition, Under-Pinning, Fire Escapes, Claddings, and Roofs.* Philadelphia: Trans-Atlantic Publications, Inc., 1987.

Fane, Francis Douglas, and Don Moore. *The Naked Warriors: The Story of the U.S. Navy's Frogmen.* Annapolis, MD: Naval Institute Press, 1995.

Halberstadt, Hans. *Demolition Equipment.* Osceola, WI: Motorbooks International, 1996.

Kennedy, Robert. *Life as an Army Demolition Expert.* New York: Children's Press, 2000.

Liss, Helene. *Demolition: The Art of Demolishing, Dismantling, Imploding, Toppling, and Razing.* New York: Black Dog & Leventhal Publishers, 2000.

Index

Demolition Experts: Life Blowing Things Up

About the Author

Mark Beyer has lived around tall buildings in Chicago and New York City. His fascination with explosions and collapsing buildings helped him write this book.

Photo Credits

Cover © AP Photo Worldwide; p. 6 © Thayer Syme/FPG; p. 10 © Corbis; p. 14 © Len Rubenstein/Index Stock; p. 17 © Kevin R. Morris/Corbis; p. 18 © Josh Mitchell/Index Stock; pp. 20, 36, 44 © AP Photo World Wide; p. 26 © AP/*The Daily Oklahoman*; p. 30 © John McNulty/Corbis; p. 33 Roger Ressmeyer/Corbis; pp. 42–43, 49, 50, 52, 53 © Reuters New Media Inc./Corbis; p. 46 © Kevin Beebe/Index Stock.

Design and Layout

Les Kanturek